Work 101

钱生钱

Making Money on Money

Gunter Pauli

[比] 冈特·鲍利　著

[哥伦] 凯瑟琳娜·巴赫　绘

姚晨辉　译

上海远东出版社

丛书编委会

主　任：田成川

副主任：何家振　闫世东　林　玉

委　员：李原原　翟致信　靳增江　史国鹏　梁雅丽

　　　　任泽林　陈　卫　薛　梅　王　岢　郑循如

　　　　彭　勇　王梦雨

特别感谢以下热心人士对童书工作的支持：

匡志强　宋小华　解　东　厉　云　李　婧　庞英元

李　阳　刘　丹　冯家宝　熊彩虹　罗淑怡　旷　婉

杨　荣　刘学振　何圣霖　廖清州　谭燕宁　王　征

李　杰　韦小宏　欧　亮　陈强林　陈　果　寿颖慧

罗　佳　傅　俊　白永喆　戴　虹

目录

Contents

一名十多岁的男孩和他的爷爷坐在一起，看着满月挂上中天。"该睡觉了。"爷爷说，"明天你们都还要早起。"

"爷爷，为什么我爸爸每天早上都必须那么早去上班？"男孩问，"他几乎没有时间吃早餐。"

"你爸爸必须辛苦工作，赚到足够的钱，这样你才有饭吃，有房子住啊。"

"为什么我妈妈也要那么早去上班？"男孩疑惑地问道。

A teenage boy sits with his granddad. They are watching the full moon rise. "Time for bed," says Granddad. "You all have to get up early tomorrow."

"Why does my dad have to go to work so early every morning, Granddad?" the boy asks. "He hardly has time for breakfast."

"Your dad has to work long hours to earn enough money so you have food on the table and a roof over your head."

"And why does my mom have to go to work so early too?" wonders the boy.

该睡觉了

Time for bed

付给他们更多的工资

to pay them more money

"为了挣钱买汽车，载她上
班啊，还要挣钱买汽油呢。"
"那为什么爸爸和妈妈每天回家都这么晚？"
"因为他们只有整天非常努力地工作，才能获得晋升。
而一旦得到晋升，他们又不得不更加努力工作，这
样才能证明付给他们更多的工资是公平
合理的！"

"To earn enough to pay for a car to get
her to work, and have money for fuel."
"And why do Mom and Dad get home
so late every day?"
"Because they have to work very hard all
day long to be promoted. And once they are
promoted they will have to work even harder
to show that it was justified to pay them
more money!"

"爷爷，你觉得这样有意义吗？而且，为什么爸爸现在要三天两头到国外旅行？"

"嗯，因为他升职了，他现在还需要负责其他国家的公司。"

"为什么妈妈总是给我们买用糖、牛奶和小麦制作的便宜食物？"

"因为他们需要省下钱，去偿还他们的债务。"

"Does that make sense to you, Granddad? And also, why is Dad travelling overseas so often now?"

"Well, as he has been promoted, he now needs to take care of business in other countries as well."

"Why is Mom always buying us that cheap food that is full of sugar, milk and wheat?"

"Because they need to save money to pay all their debt."

省下钱去偿还他们的债务

save money to pay all their debt

现在就享受美好生活，以后再为这一切买单

Live well now, pay for everything later

"可是，为什么他们一开始就要欠下这么多债务呢？"

"因为你的父母想现在就享受美好生活，以后再为这一切买单。"

"但是，为了这些债务，他们几乎从来不在家！他们不得不整天工作，因为最终他们将付出更多的钱还债。即使他们回到家，也因为太累而不能陪我！所以，我的父母就像是在坐牢一样！"

"But why do they have so much debt in the first place?"

"Because your parents want to live well now. And pay for everything later."

"But with all their debt, and having to work so much harder because it costs them more in the end, they are never at home! And when they are home, they are too tired to spend time with me! So my parents are virtually in prison!"

"不，我的孩子，不要这么想。

你的父母都是自由的！他们自愿作出这

种选择，是为了让你生活得更好。他们爱你，希望

你拥有最好的生活。"

"我知道他们爱我，我也真的很感激他们为我做的一切。"

"你的父母确实有债务，他们需要支付利息，但这样就可

以满足你的所有需要，让你顺利成长，拥有一个光

明的未来。"

"No, my boy. Don't think of it like that.
Your parents are free! Free to make choices
that will enhance your life. They love you and
want you to have the best."

"I know they love me, and I am really
grateful for all they do for me."

"Your parents do have debt and they do pay
interest, but that is so you can grow up with
all you need and have a bright future."

真的很感激他们为我做的一切

really greatful for all they do for me

13

没有办法再供养房子和汽车

拍　卖

Can no longer pay for the house and the car

"我看得出来，爷爷。不过，如果有一天，他们工作的公司不需要他们了，会发生什么呢？"

"哦，他们会找到其他工作……我们希望是这样。"爷爷说。

"但是，如果他们不能马上找到工作，没有办法再供养房子和汽车，又会怎样呢？"

"I see that, Granddad. But what happens if one day the company they work for does not need them anymore?"

"Oh, they will find other jobs... we hope," says Granddad.

"But what if they don't right away and can no longer pay for the house and the car?"

"他们将会用他们的房子作抵押，借更多的钱，让我们渡过难关，直到他们找到别的工作。"

"我不明白，谁在通过发放贷款赚钱呢？"

"是银行，你向银行借钱，银行借给你钱，这样你可以生活下去，花钱买房子、汽车和其他的东西。"

"Then they will borrow even more money, using their house as a guarantee, to see us through until they find other jobs."

"I don't understand. Who is making money from lending people money?"

"The bank, you borrow from the bank. The bank lends you money so you can progress in life, buy a house, a car, and other things that cost a lot."

Who is making money from lending people money?

钱是造币厂印制的。

The mint does that.

"那么银行的

钱是从哪里来的？难道是他

们印出来的吗？"

"不是，钱是造币厂印制的。我给你解释一下银行是

怎么工作的：那些有钱却暂时不想花的人会将钱存在银

行的储蓄账户中。"

"那么，那些很有钱的人不花光他们所有

的钱吗？"

"And where does the bank get all
the money from? Do they print it?"

"No, the mint does that. To explain how the
bank works: People who have more money
than they need right now put that money in a
savings account at the bank."

"So people who have lots of money don't
spend all their money?"

"是的。他们把钱存
在银行里。银行用这笔钱来赚
钱。将钱存在银行储蓄账户的人也用自己的
钱赚了一些钱，因为银行为了留住这些钱，会付给
他们一点儿利息。"

"人们怎么能够通过钱来赚钱？这就像用水来制作水一
样！这是不可能的。我不明白这样的情形长期存在的
道理，你知道吗，爷爷？"

……这仅仅是开始！……

"No, they keep it in the bank.
The bank uses that money and
makes money from it that way. People
who keep their money in savings accounts
at the bank also make some money on their
money as the bank pays them a bit of interest
on it, to keep it there."

"How can one make money from money? It's
like making water from water! It is impossible.
I can't see it lasting forever, can you
Granddad?"

... AND IT HAS ONLY JUST
BEGUN!...

……这仅仅是开始！……

··· AND IT HAS ONLY JUST BEGUN! ···

When debt increases faster than earnings, due to higher interest rates or the need to borrow more, then consumption must decrease, and poverty rises.

由于利率较高，或者需要借更多的钱，导致债务增长快于收入时，消费水平就必定会降低，而贫困程度就上升了。

People are not the only ones taking on debt; governments borrow a lot of money as well. If government debt increases, then taxes must be increased, and this leaves less money for families and companies to buy or to invest.

并不是只有个人才会负债，政府有时也会借很多的钱。如果政府的债务增加，那么就必须增加税收，家庭和企业用于购买或投资的钱就会变少。

When families have more debt, their creditworthiness decreases, meaning that the banks consider them a higher risk, this means that they will have to pay higher interest rates, which decreases their capacity to pay back the loan.

当家庭债务增加时，其资信会降低，这意味着银行认为他们的风险较高，他们将不得不承担更高的利率，从而降低他们偿还贷款的能力。

The total world debt is US$ 230 trillion, which is 3 times more than annual GDP (gross domestic product).

全球的总债务为230万亿美元，比全年GDP（国内生产总值）的3倍还要多。

Debt leads to decreasing economic growth. In the year 2000 for every US$ 2.4 of debt creation there was US$ 1 added to GDP. By 2015 it was necessary to accept US$ 4.6 in additional debt to create $1 extra in GDP.

债务导致经济增长减缓。2000年，每2.4美元的债务会使GDP增加1美元。到2015年，为了使GDP增加1美元，人们不得不接受4.6美元的额外债务。

The US alone pays US$ 400 billion in interest per year, while the rates are at a historically low level.

美国每年要支付4000亿美元的利息，这还是利率处于较低历史水平的情况下。

$ 400亿
p/y

In the
industrialised world,
the greatest debt is held by
households and companies (41%),
and in the emerging economies
the greatest debt is held by the
financial sector (43%).

在工业化国家，最大的债务由家庭和企业持有（41％），而在新兴经济体国家，最大的债务由财政部门持有（43％）。

Australian
households are the
most indebted in the world.
Their total loans from banks
represents 130% of the total amount
of products and services produced
each year. This makes Australians
very vulnerable to any change
in interest rates.

澳大利亚的家庭是世界上负债最重的，他们的银行贷款总额是其每年产品和服务总额的130％，这使得澳大利亚人非常容易受到利率变化的影响。

Think about It

想一想

Would you like to work to pay off debt, or would you like to work and save money and in this way be able to afford buying something without having to become indebted?

你是愿意通过工作来偿还债务，还是愿意先工作攒钱再购买自己需要的东西，而不必背负债务？

你希望父母花更多的时间陪伴你，还是希望他们工作更长时间来赚更多的钱？

Would you like your parents to be spending more time with you or would you like them to work longer hours to make more money?

Do you approve of your parents saving money by buying cheap and unhealthy food, or would you like them to save by taking public transport instead of driving their own cars and buying healthy food?

你赞成你的父母通过购买廉价而不健康的食品来省钱，还是希望他们不再开车，通过改乘公共交通工具却购买健康食品来省钱？

你认为过度负债就像是在坐牢一样吗？还是说，你将这视作一种享受生活的好机会？

Do you consider excessive debt to be like in prison? Or do you see it is a great opportunity to enjoy life?

Do It Yourself!

自己动手!

Do you receive pocket money from your parents for doing chores such as making your bed, doing the dishes, mowing the lawn, walking the dog, taking the garbage out, washing their cars, doing some shopping or perhaps even cooking a meal? Propose a plan to your parents whereby you will help with such chores and have them deposit the money you have earned in a separate savings account for you. How much money do you think will you have after ten years?

你会通过做家务从父母那里获得零用钱吗？比如说整理自己的床、洗餐具、修剪草坪、遛狗、倒垃圾、清洗父母的汽车、购物或者做饭。向你的父母提出一个计划，你会帮助他们做这类家务，让他们为你开设单独的储蓄账户，将你赚的钱存进去。你认为自己在10年后会攒下多少钱？

学科知识
Academic Knowledge

生物学	食物作为所有生命的基本需要及其生产；从浆果、水果、蔬菜和肉类到牛奶和小麦，这些食物的消化难度是递增的。
化 学	摄入的糖、牛奶和小麦会在体内进行不同的新陈代谢，并产生副作用。
物 理	地球上只有这么多水，你可以改变它，但永远不能创造更多的水。
工程学	金融工程是一门新兴学科，它使用复杂的金融工具，以构架财务和风险。
经济学	利率和风险之间的平衡；偿还债务的能力取决于创收的能力；获得贷款需要提供担保；抵押贷款工具；银行"创造"钱的能力在于放贷比存款多；中央银行作为最后贷款人的作用；新兴的小额信贷体系；信用合作社（目的是为老百姓服务）和银行（目的是为股东赚钱）之间的差异；新兴的网上银行和大众融资的作用。
伦理学	放高利贷的人索要过高的利率，并要求全额担保，这使得一些借款人陷入绝望，导致他们自杀；缺乏职业道德地提供更多贷款会带来更多无法偿还的债务；政府的作用是在发生自然灾害（不可抗力）时进行干预；在2008年金融危机之后，政府拯救了贷款机构，但没有顾及私人公司或家庭；助学贷款债务危机，其总额超过了1万亿美元；高频交易能利用利率、汇率和股价的微小差异获利数十亿美元，在一瞬间买进和卖出数百万次；整个经济体系的目的是鼓励巨额债务。
历 史	注销债务和还款宽限期的引入首次出现在3 500年前的美索不达米亚；美国政府于1819年强制推行了债务延期偿还，以拯救歉收的农民；在1929年大萧条时期，农民被放任其破产。
地 理	伦敦和纽约是世界主要的金融中心；月球运行周期与月亮的盈亏。
数 学	金融数学即计量金融学，把数学及数值模型衍生和扩展至金融市场；由超级计算机支持的高频交易，几秒钟可以执行数千次交易。
生活方式	借债的习惯；随着时间的推移支付更高的金额，包括利息，然后用收入偿还，提供担保；不同于土地、劳动和贸易，金钱和债务是虚拟商品。
社会学	现代社会是建立在债务基础之上的，这并不一定会导致团结一致；法律的经济社会学把关于债务关系的法律、经济和社会学关联在了一起；债务关系是跨学科的；金钱使人堕落的信念；借和贷之间的区别。
心理学	负债的决定不仅仅取决于你的资产，更主要的是你的自信程度；负债过多的人会感到孤立、内疚和惭愧，可能会导致节制；高负债会妨碍最佳运作；债务带来的压力会影响一个人的幸福感。
系统论	金钱使世界运转，但过度负债会让世界停摆。

情感智慧
Emotional Intelligence

孙 子

小孙子注意到爸爸需要很早去上班，往往不能与家人一起吃早餐。他在晚上和饭桌上想念父母，对爸爸经常长时间出国旅行的事实感到焦虑。他很奇怪妈妈为什么会购买廉价却不健康的食物，并对他的家庭负有巨额债务的事实感到压力重重。他认为债务就像是一所监狱，因为它限制了他父母的自由，让他们没办法花更多的时间陪伴他。他意识到父母将来可能面临无法偿还债务的风险。孙子对攒钱的方式提出了质疑，并且看清了这一切已经失控的事实。

爷 爷

爷爷费力地解释赚钱的必要性。他很心疼自己的孙子，一步一步地解释为什么他的父母过着现在的生活，不能像孩子渴望的那样花更多的时间陪伴他，并展示了清晰的逻辑性。他相信一切都会好起来，并且孙子也能够理解和接受这一切。爷爷听到监狱的比喻吃了一惊，坚持认为父母都是自由的，他们之所以做这样的选择是出于他们对儿子的爱。当孙子提到父母失去工作的风险时，爷爷仍保持了坚定的信仰，并提供了一个临时解决方案：借更多的钱！爷爷精准地解释了银行是如何运作的，并用通俗易懂的语言分享了他的见解，受到这一逻辑触动，孙子有了进一步思考。

艺术
The Arts

过度负债可能导致愤怒的情绪。涂鸦往往用来表达愤怒。你准备好测试自己创造涂鸦的能力了吗？征得你的父母同意后购买喷漆，要确保你买的是水溶性油漆。说服你的父母或老师允许你在家里或学校用一堵墙进行涂鸦。选择一个事后容易清洗的地方。现在，挑选一种你喜欢的风格设计你的涂鸦，并与大人们商量你准备使用的词语。房间里是否有一名专家乐意演示如何涂鸦呢？

思维拓展
Systems: Making the Connections

　　俗话说，金钱让世界运转。借钱是一项存在了几千年的经济活动，私人和公共机构的债务积累已经达到令人难以理解的程度。这个世界必须向金融机构支付三倍于国内生产总值债款的事实，证明了我们处于过度消费和过度借贷的状态，这最终将造成对地球的过度开发。即使世界每年保留其生产总值的10％以抵消部分债务，仍然需要不止一代人的努力才能使债务回到合理水平。造成这一切的原因是利息。当利率较低时，形势是可控的，一旦利率上升到一定程度，许多家庭和政府的财政将会崩溃。但即使远在这样的崩溃之前，利率也会导致收入的不公平分配，穷人会变得更穷，富人会变得更富。利息负担主要落在贫困人口身上，使他们无法获得社会或经济的发展。债务清偿减少了提供社会服务、健康和营养的能力，将导致人们选择劣质廉价的产品。这种逻辑同样适用于一个国家，特别是发展中国家。我们也可以用投资的逻辑来取代严格的贷款法律原则。如果有人有兴趣帮助一个家庭或一个国家成长和发展，也可以不是出借本金以赚取利息，而是签订一个投资协议，双方共享产出和收益。当今的逻辑是借出钱的人会得到一个固定回报，而借钱者拥有不确定的回报。这意味着风险完全落在了借款人的肩上，而希望用钱来赚钱的贷款人没有任何风险。这种做法已成为当前的标准模式，但还有没有另外的选择？关键的问题是：现在改变是否为时已晚？

动手能力
Capacity to Implement

　　如果你还没有开始攒钱，现在行动起来吧。找些家务做做，不要将你赚到的钱花掉，把它存起来，看看你能积攒多少。问问你的朋友们，谁愿意共同建立一个特殊账户，每个人都可以根据他们赚钱的手段和能力作出贡献。当你已经积攒了一定金额，问自己一个问题："为了给所有人谋福利，我们会一致同意进行什么投资？"创建你们自己的小额信贷联盟，如果你们选择的项目是成功的，约定拿出部分利润来增强你们的"资金池"。通过这种方式，你们就积累了社会资本，而不是将它们花掉和产生债务。

故事灵感来自
This Fable Is Inspired by

穆罕默德·尤努斯
Muhammad Yunus

穆罕默德·尤努斯出生于孟加拉国一个穆斯林家庭，在九个孩子中排行第三。他是一名活跃的理想主义者，并且擅长戏剧。在完成学业之后，穆罕默德被任命为孟加拉国吉大港大学的讲师，后在美国范德比尔特大学获得经济学博士学位。意识到自己国家的贫穷和饥饿水平以及高利贷的习俗后，他与阿赫塔·哈米德·汗（Akhtar Hameed Khan）博士一起构思了小额贷款的想法，即以合理的利率进行非常小额的贷款。他发起的乡村银行（Grameen Bank）和小额贷款的成功，启发了100多个国家的类似努力。2006年，穆罕默德·尤努斯和乡村银行被联合授予了诺贝尔和平奖。

图书在版编目（CIP）数据

冈特生态童书.第三辑修订版：全36册：汉英对照 /
（比）冈特·鲍利著；（哥伦）凯瑟琳娜·巴赫绘；
何家振等译.—上海：上海远东出版社，2022
书名原文：Gunter's Fables
ISBN 978-7-5476-1850-9

Ⅰ.①冈… Ⅱ.①冈… ②凯… ③何… Ⅲ.①生态环
境–环境保护–儿童读物—汉、英 Ⅳ.①X171.1-49

中国版本图书馆CIP数据核字（2022）第163904号
著作权合同登记号图字09-2022-0637号

策　　划　张　蓉
责任编辑　程云琦
封面设计　魏　来　李　廉

冈特生态童书
钱生钱
[比]冈特·鲍利　著
[哥伦]凯瑟琳娜·巴赫　绘

姚晨辉　译

记得要和身边的小朋友分享环保知识哦！
八喜冰淇淋祝你成为环保小使者！